I0824267

Microscope

Adrienne Beaucage

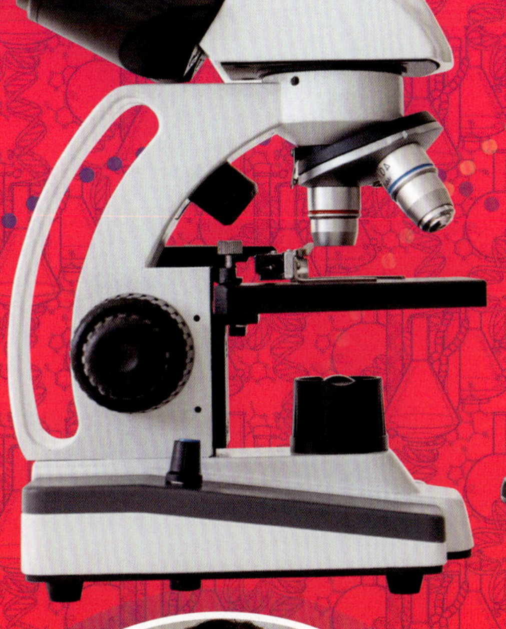

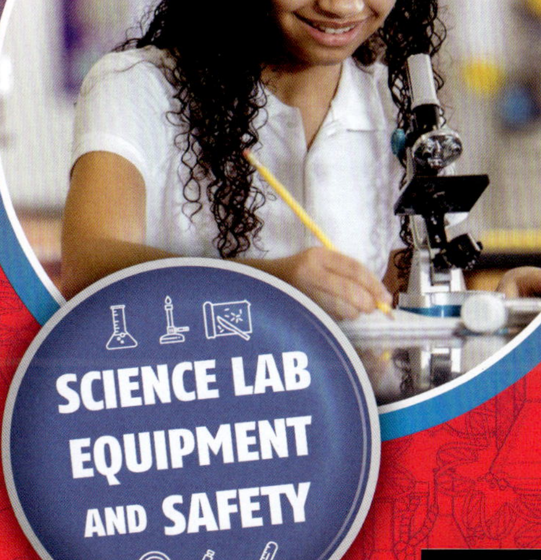

SCIENCE LAB EQUIPMENT AND SAFETY

LIGHTBOX
openlightbox.com

Go to
www.openlightbox.com
and enter this book's
unique code.

ACCESS CODE

LBXW5724

Lightbox is an all-inclusive digital solution for the teaching and learning of curriculum topics in an original, groundbreaking way. Lightbox is based on National Curriculum Standards.

LIGHTBOX SUPPLEMENTARY RESOURCES

SHARE
Share titles within your Learning Management System (LMS) or Library Circulation System

CURRICULUM
Find national and state curriculum correlations

CITATION
Create bibliographical references following the Chicago Manual of Style

STANDARD FEATURES OF LIGHTBOX

AUDIO High-quality narration using text-to-speech system

ACTIVITIES Printable PDFs that can be emailed and graded

SLIDESHOWS Pictorial overviews of key concepts

VIDEOS Embedded high-definition video clips

WEBLINKS Curated links to external, child-safe resources

TRANSPARENCIES Step-by-step layering of maps, diagrams, charts, and timelines

INTERACTIVE MAPS Interactive maps and aerial satellite imagery

QUIZZES Ten multiple-choice questions that are automatically graded and emailed for teacher assessment

KEY WORDS Matching key concepts to their definitions

This title is part of our Lightbox digital subscription

Lightbox Grades 3–5 Subscription
ISBN 978-1-5105-5424-5

Access hundreds of Lightbox titles with our digital subscription. Sign up for a **FREE** subscription trial at **www.openlightbox.com/trial**

Microscope

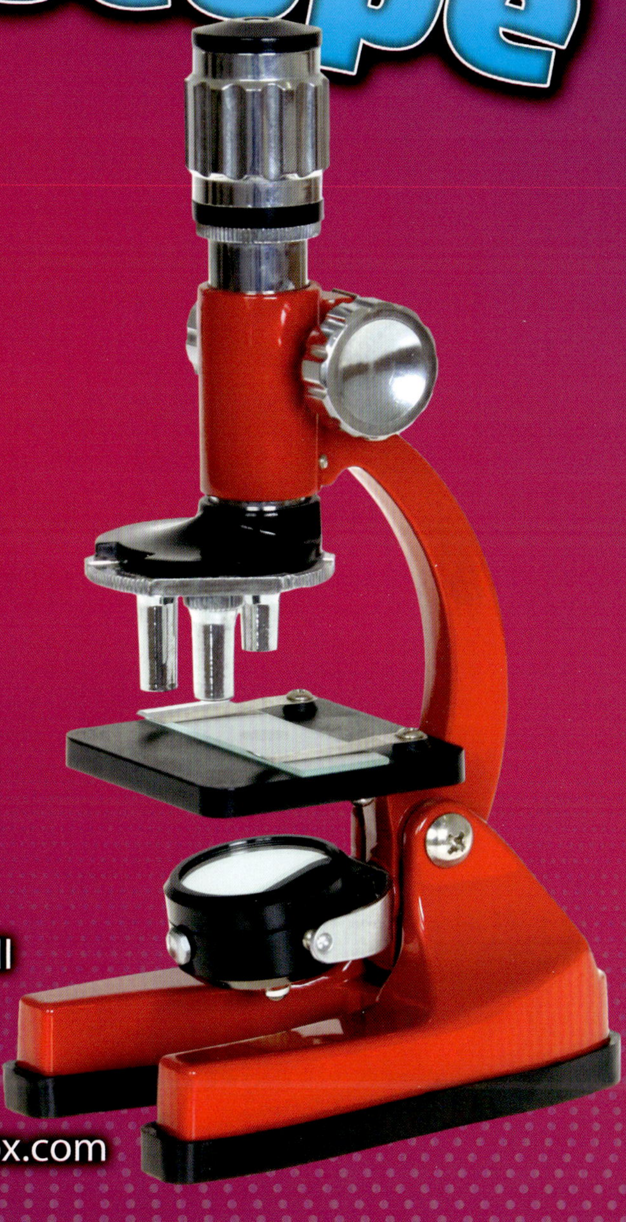

CONTENTS

A CLOSER LOOK

Have you ever been in a science laboratory, or lab, before? There might be one in your school. Science labs have special tools and equipment that can be used to conduct **experiments**. Scientists perform experiments to discover more about the world. You can conduct experiments in a lab to learn new things, too.

Microscopes are a common piece of equipment found in science labs. There are many types of microscopes that work in different ways. However, all microscopes have one thing in common. They **magnify** very small things. This helps scientists **study** objects that would otherwise be too small to see. Using a microscope can be a window into parts of the world that are not usually visible to you.

The world's **most powerful** microscopes are more than **12 feet** (4 meters) **tall**.

The name **"microscope"** comes from **two Greek words**. ***Mikros*** means "small," and ***skopein*** means "to look at."

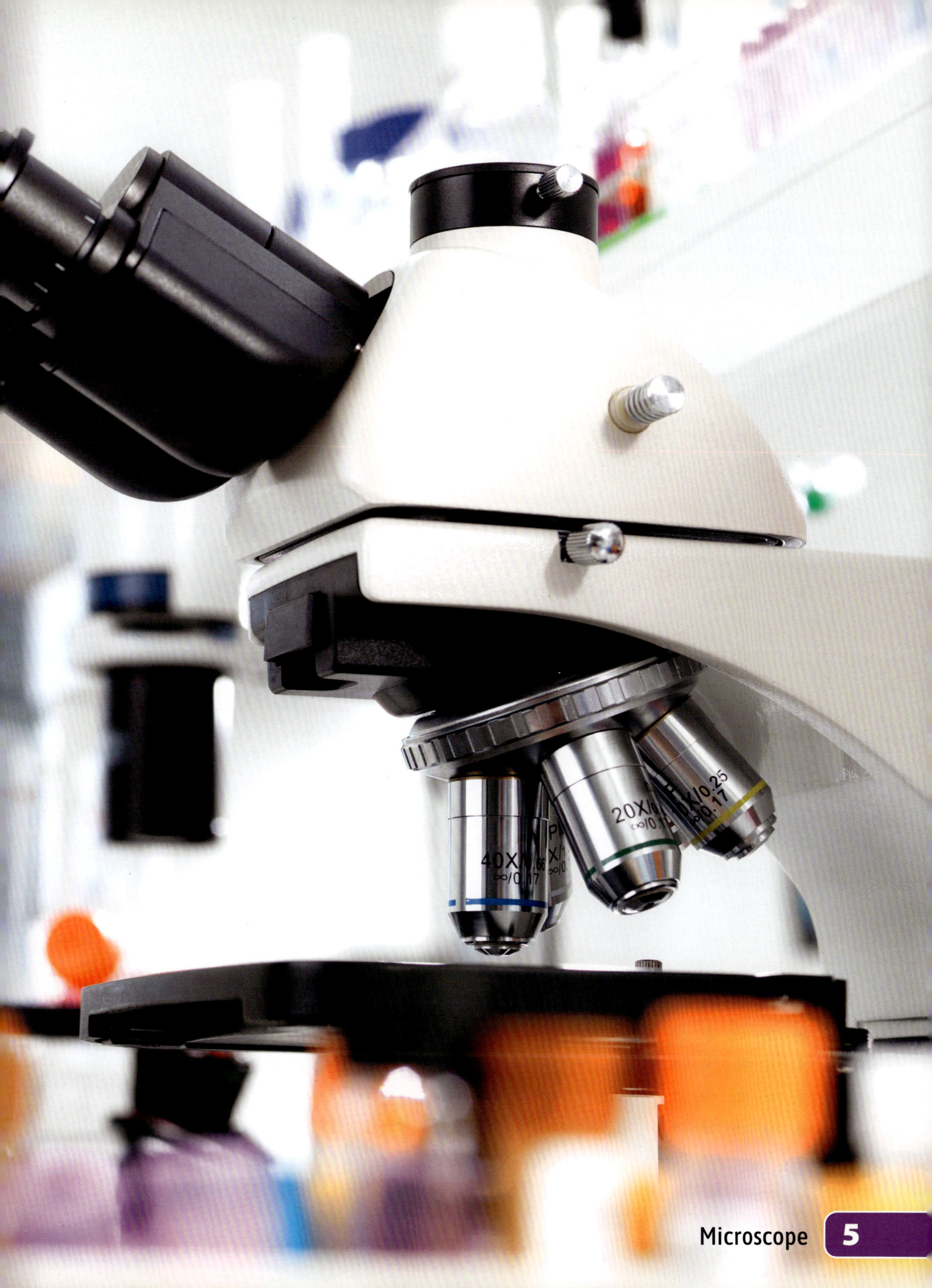
40X/0.66
∞/0.17
20X/0.
∞/0.1
10X/0.25
∞/0.17

MICROSCOPE HISTORY

People have used tools to magnify objects for centuries, but the first **compound microscope** was invented in about 1590. Hans Janssen and his son, Zacharias, are often credited with the invention. Hans made **spectacles**. He and Zacharias realized that they could use a tube with a lens at the top and bottom to magnify objects. This device was a step toward the development of future microscopes.

Hans and Zacharias Janssen's microscope could magnify an object up to nine times its actual size.

Over the next few hundred years, many scientists worked to modify the microscope. Using multiple and better lenses made it more powerful. As the device improved, it helped scientists understand more about **cells** and **biology**. Several scientific discoveries would not have been possible without the invention of the microscope.

Microscope Timeline

1665
A scientist named Robert Hooke publishes a book called *Micrographia*. It is full of pictures he drew of things he saw through a microscope.

1931
The **electron microscope** is invented by two scientists, Ernst Ruska and Max Knoll.

2020
Scientists invent a new type of microscope that can be used to see straight through a skull.

Eyepiece
Lens used to look through the microscope

PARTS OF A MICROSCOPE

A microscope has many different parts that work together to magnify objects. The object being examined, called the specimen, is put on a glass **slide**. This is placed on the microscope's stage. A light source is used to **illuminate** the slide, making it easy to see the specimen through the eyepiece. The microscope's lenses are made from a special kind of glass. They magnify the specimen.

Stage
Where the slide with the specimen is placed

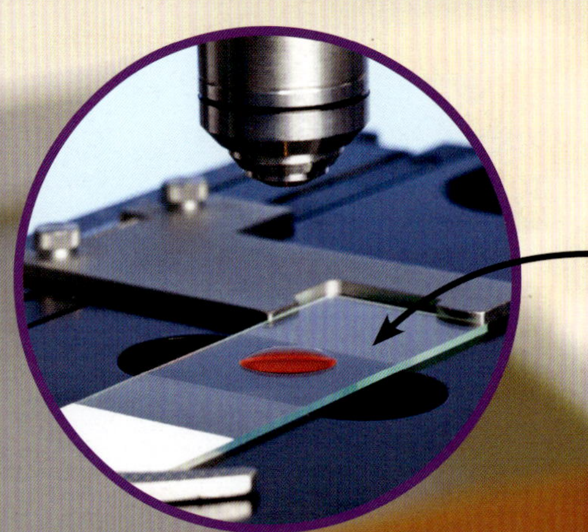

Glass slide

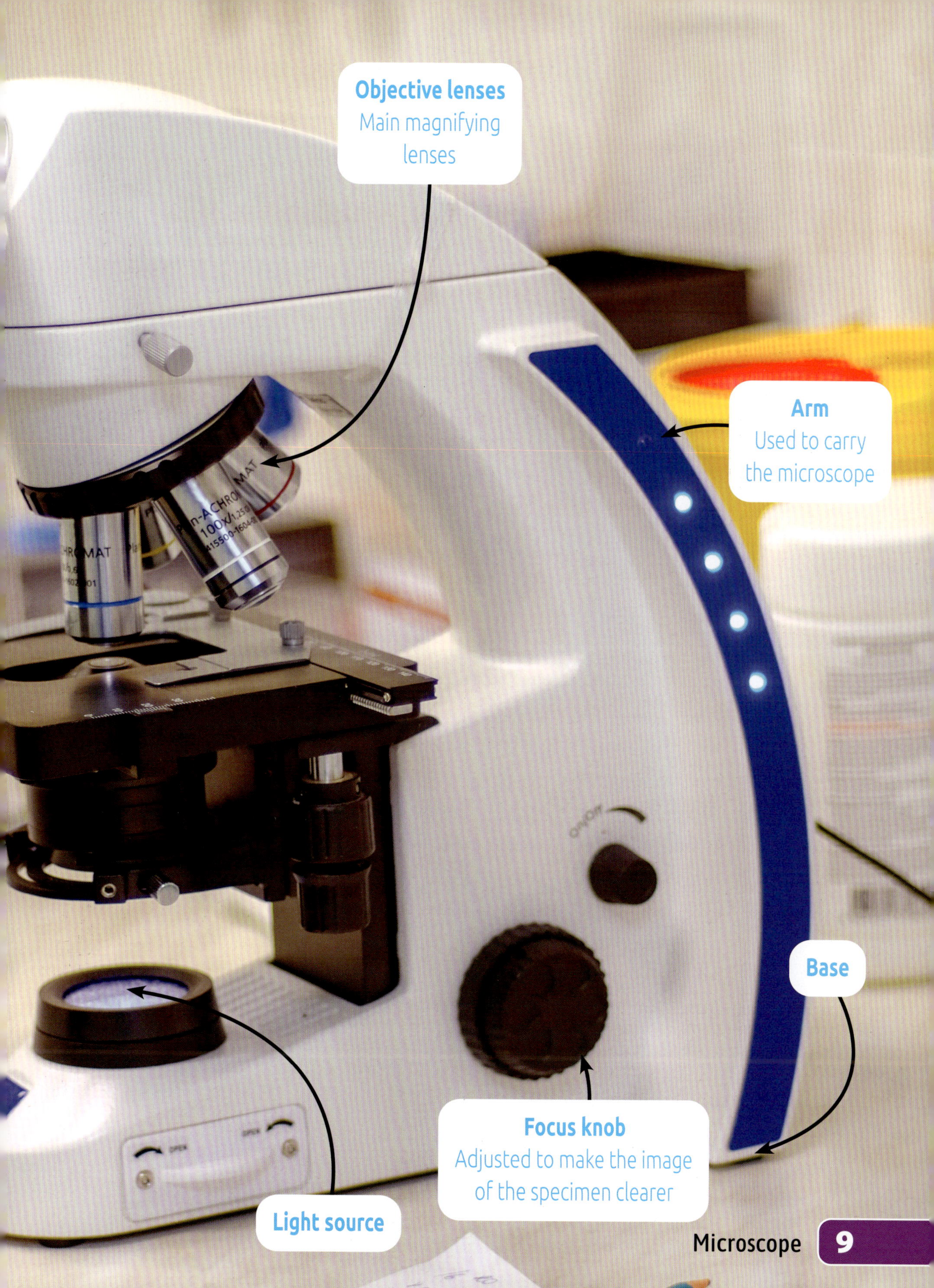
Objective lenses
Main magnifying lenses
Arm
Used to carry the microscope
Base
Focus knob
Adjusted to make the image of the specimen clearer
Light source

HOW IT WORKS

The most common type of microscope found in school science labs is a light microscope. Sometimes called an optical microscope, it works like a magnifying glass. The specimen sits underneath a lens, which is a curved piece of glass. The curved glass causes light to bend. As light passes from the specimen through the lens, the specimen is magnified and appears larger.

Most school microscopes magnify objects by 10 to 400 times.

A compound microscope is a type of light microscope. It has multiple lenses. The objective lens is close to the object, while the other lens, the eyepiece, is farther away. When light travels through the first lens, the specimen is magnified. Then, the light travels through the eyepiece and bends again. This makes the specimen look even larger.

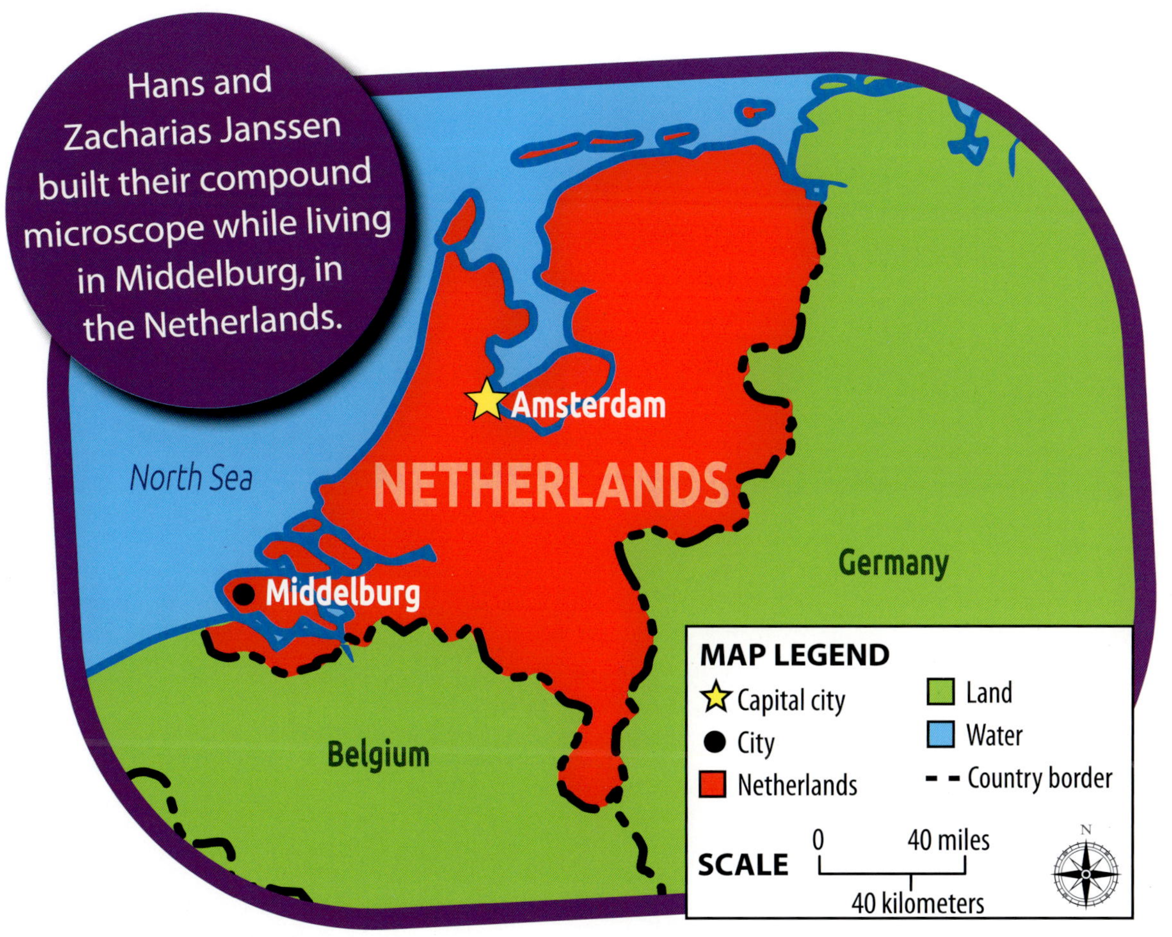

SAFETY IN THE LAB

Always place a microscope on a flat surface so it will not slide or fall.

It can be exciting to learn new things using the special equipment found in a science lab. However, it is important to learn the safety rules before using any equipment. Certain objects and tools can be dangerous if they are not used properly.

A microscope is an expensive piece of equipment that can easily be damaged if it falls. Always use two hands to carry a microscope. Hold the arm of the microscope with one hand, and place the other hand underneath the base. Do not run with the microscope or swing it around. Be careful not to hit it against any chairs or tables. If there is any broken glass in or around a microscope, do not touch it. Stop the experiment and tell a teacher right away.

An object viewed through a **compound microscope** can appear up to **2,000 times larger** than it actually is.

A **human hair** is about **80,000 nanometers** wide. Light microscopes can be used to see objects as small as **500 nanometers** wide.

THE SCIENTIFIC METHOD

Scientists follow a certain process called the scientific method. It is important to learn it and follow the steps correctly. Using this method improves the results of an experiment.

1. QUESTION

Think of a question that you would like to answer.

2. RESEARCH

Learn more about the topic. You can use books or online research to help you gather information.

3. HYPOTHESIS

Use your research to help you predict, or guess, the answer to your question. Write down your guess. This is called a hypothesis.

4. EXPERIMENT

Design an experiment to test your hypothesis.

5. OBSERVATION

What happened during the experiment? Observe the results and write down what you see.

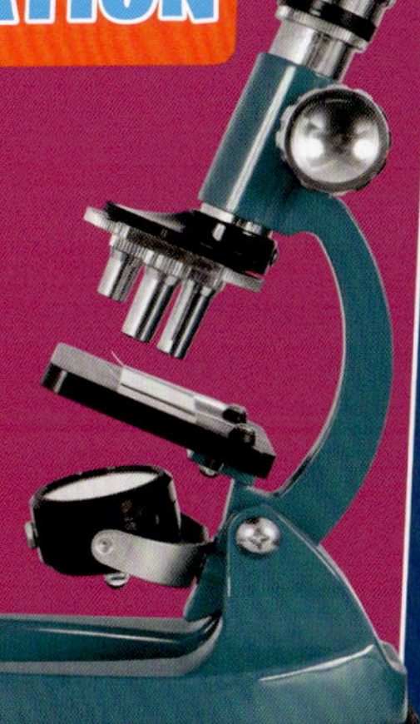

6. ANALYSIS

Review your observations and analyze the results. Can you determine what happened and why? The result, or thing that happened, is called the effect. The reason for the effect is the cause.

7. CONCLUSION

Was your original hypothesis correct? Write about what happened and whether you accept or reject the hypothesis. Share the results of your experiment with others so they can learn, too.

EXPERIMENT

SALT OR SUGAR?

Wendy is learning about microscopes in school. Her teacher has set up two bowls filled with small white crystals. One bowl has powdered sugar in it, and the other has salt. The teacher knows which one is which, but they look the same to Wendy. Can she use a microscope to tell which is sugar and which is salt?

1. QUESTION

Is it possible to use a microscope to tell sugar and salt apart?

2. RESEARCH

To the **naked eye**, powdered sugar and salt look the same. However, Wendy knows that they do not taste the same. When she looks up how salt and sugar are formed, she learns that they are made of tiny pieces called crystals that are quite different in shape and color.

SAFETY FIRST!

Sugar and salt are foods, but many other substances used in the science lab are not. Some substances can be very dangerous to eat. Do not lick, taste, or swallow any substance in a science lab.

3. HYPOTHESIS

Sugar and salt crystals will look different from each other when viewed under a microscope.

4. EXPERIMENT

Wendy scoops a small spoonful of one substance from its bowl and puts it on her microscope slide. She takes a look through the microscope and notices the color and shape of the crystals. Next, she repeats the process with a small spoonful of the other substance.

5. OBSERVATION

Under the microscope, the crystals of the first substance look like tiny cubes. They are a frosted white color. The crystals of the second substance look more like **hexagons** and are clear.

6. ANALYSIS

Since microscopes magnify objects, the crystals appear larger. This makes it possible to see greater detail than is visible to the naked eye. Wendy knows from her research that sugar crystals are clear and hexagonal, while salt crystals are white and cubic.

7. CONCLUSION

The experiment supported Wendy's hypothesis. Seeing objects in detail under a microscope makes it possible to tell sugar and salt apart, even if they appear the same to the naked eye.

EXPERIMENT

SAFE TO DRINK?

Jeff is curious about the pond behind his house. His mother has told him not to drink from it, even though the water looks clean and clear. Jeff wonders why. He thinks there could be **organisms** in the water that are too small for him to see.

1. QUESTION If water looks clear, is it safe to drink?

2. RESEARCH Jeff visits the library at his school to do some research. He learns that many small organisms can be found in pond water, including **bacteria**, which are not visible to the naked eye. Many bacteria are dangerous.

SAFETY FIRST!

Even if water looks clear under a microscope, do not drink it. Wash your hands using soap and clean water after working with samples in a science lab.

3. HYPOTHESIS Pond water contains organisms that are too small to see with the naked eye.

4. EXPERIMENT Jeff's mother helps him collect a **sample** of pond water in a small jar. He uses a dropper to take water from the jar and drop it onto a glass slide. Jeff then places the slide on his microscope's stage and looks at the water through the lens.

5. OBSERVATION Jeff is patient while he observes the water sample. He sees a small creature moving across the slide. He sketches a picture of it in his notebook.

6. ANALYSIS The water in the pond contains small living organisms, making it unsafe to drink.

7. CONCLUSION The experiment supported Jeff's hypothesis. Pond water that looks clear may contain tiny organisms.

Water that is safe to drink is called drinking water. It is treated to remove harmful organisms.

ACTIVITY

EXAMINE A DOLLAR BILL

Money has many tiny features that are hard to see with the naked eye. Follow these steps to examine what a dollar bill looks like under a microscope.

MATERIALS

- Microscope
- Dollar bill

STEPS

1. Place the dollar bill on your microscope stage. The bill is flat, so you do not need to use a slide.

2. Adjust your microscope settings to allow as much light as possible to shine from under the bill.

3. Look at the bill through the microscope.

4. Slowly move the bill around the stage.

5. Do you notice any shapes or colors? You may be able to see that the bill actually has thin red and blue threads running through it.

6. Can you find anything else when looking at the bill through the microscope? Write down your findings and share them with your classmates.

HELPFUL HINT

Money can spread germs when it is passed from person to person. It is important to wash your hands with soap and water after touching money.

QUIZ

1 Is there more than one type of microscope?

2 How tall are the world's most powerful microscopes?

3 When was the first microscope invented?

4 Who is credited with inventing the first compound microscope?

5 In what year was the electron microscope invented?

6 What is another name for a light microscope?

7 Which part of a microscope causes light to bend?

8 What color are salt crystals?

9 Is clear pond water safe to drink?

10 What does the Greek word *mikros* mean?

Answers:
1. Yes, there are many types of microscopes **2.** More than 12 feet (4 m) tall **3.** About 1590 **4.** Hans and Zacharias Janssen **5.** 1931 **6.** Optical microscope **7.** The curved glass of the lens **8.** White **9.** No **10.** "Small"

KEY WORDS

bacteria: tiny, single-celled organisms that can be found in all natural environments

biology: the study of life and living things

cells: the small units that living things are made of

compound microscope: a type of microscope that sits upright and uses multiple lenses

electron microscope: a type of microscope that uses a beam of tiny particles called electrons to view an object instead of using light

experiments: scientific procedures performed to discover something

hexagons: six-sided shapes

illuminate: light up

magnify: make something appear larger than it is

naked eye: viewing an object without a microscope, telescope, or any other device

organisms: living things, such as plants or animals

sample: a small amount of a substance taken for analysis

slide: a thin, flat piece of glass used to hold an object for viewing under a microscope

spectacles: glasses

study: look closely at something to observe or learn about it

INDEX

LIGHTBOX

SUPPLEMENTARY RESOURCES

Click on the plus icon found in the bottom left corner of each spread to open additional teacher resources.

- Download and print the book's quizzes and activities
- Access curriculum correlations
- Explore additional web applications that enhance the Lightbox experience

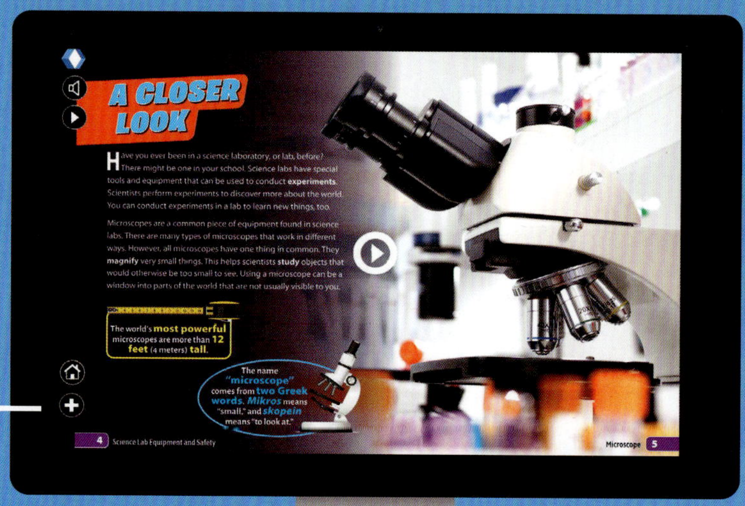

LIGHTBOX DIGITAL TITLES
Packed full of integrated media

VIDEOS

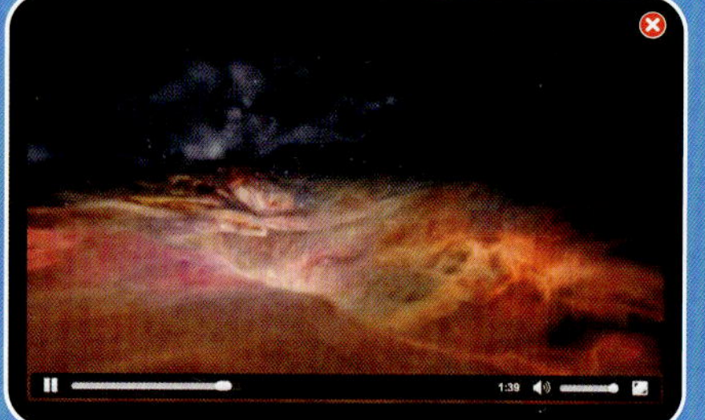

INTERACTIVE MAPS

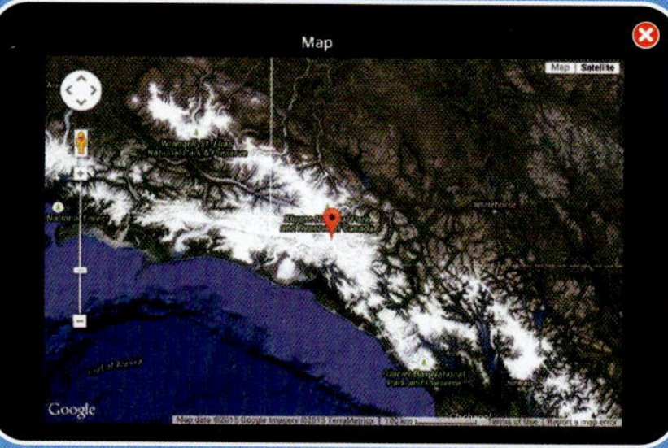

WEBLINKS

SLIDESHOWS

QUIZZES

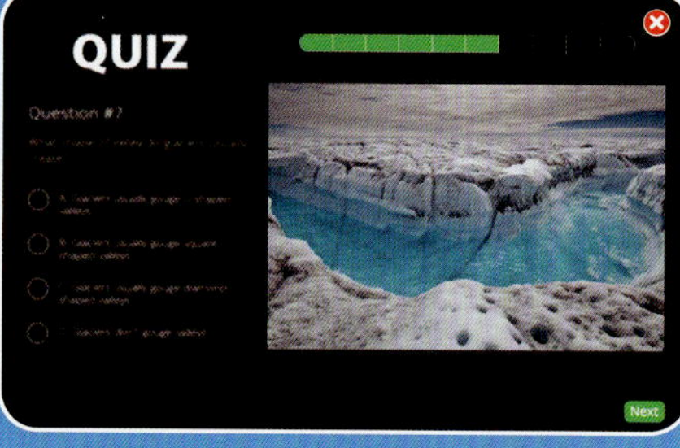

OPTIMIZED FOR

- ✔ TABLETS
- ✔ WHITEBOARDS
- ✔ COMPUTERS
- ✔ AND MUCH MORE!

Published by Lightbox Learning
276 5th Avenue, Suite 704 #917
New York, NY 10001
Website: www.openlightbox.com

Library of Congress Cataloging-in-Publication Data

Names: Beaucage, Adrienne, author.
Title: Microscope / Adrienne Beaucage.
Description: New York : Lightbox, [2022] | Series: Science lab equipment and safety | Includes index. | Audience: Grades 2-3
Identifiers: LCCN 2021032099 (print) | LCCN 2021032100 (ebook) | ISBN 9781510558991 (library binding) | ISBN 9781510559004
Subjects: LCSH: Microscopes--Juvenile literature.
Classification: LCC QH278 .B43 2022 (print) | LCC QH278 (ebook) | DDC 502.8/2--dc23
LC record available at https://lccn.loc.gov/2021032099
LC ebook record available at https://lccn.loc.gov/2021032100

Printed in Guangzhou, China
1 2 3 4 5 6 7 8 9 0 25 24 23 22 21

092021
111020

Project Coordinator: Priyanka Das **Designer:** Ana María Vidal

Every reasonable effort has been made to trace ownership and to obtain permission to reprint copyright material. The publisher would be pleased to have any errors or omissions brought to its attention so that they may be corrected in subsequent printings. The publisher acknowledges Alamy, Getty Images, and Shutterstock as its primary image suppliers for this title.